MW00377627

CW Geek's Guide to Having Fun with Morse Code

Dan Romanchik, KB6NU

Copyright © 2015 Daniel M. Romanchik

All rights reserved. No part of this publication may be reproduced, stored in a retrieval system, or transmitted in any form or by any means, electronic, mechanical, recording or otherwise, without the prior written permission of the author.

First edition

ISBN: 0692367241
ISBN-13: 978-0692367247

CONTENTS

INTRODUCTION

I got my Novice license way back in 1971, at the age of 16. Back then, Novices were restricted to operating Morse Code on small portions of 80m, 40m, 15m, and 10m. In addition, you had to upgrade within a year or lose your license. That meant either settling for a Technician Class license, or biting the bullet and passing the 13 word-per-minute (wpm) General Class code test.

Fortunately, I enjoyed operating Morse Code, and I was more than ready for the 13 wpm by the time that I'd learned enough theory to pass the written test. I took the bus down to the Federal Building in Detroit, easily passed the code test, and took the written test.

Then, the waiting began. In those days, they didn't tell you right away if you passed the test or not. You had to wait several weeks before you got a letter from the FCC with either your new license or a letter informing you that you didn't pass.

I was pretty confident that I would pass, so I'd purchased and built a Heathkit HW-101. This was one of the least expensive HF transceivers on the market at the time, costing $250 for the kit. It was capable of operating both single-sideband (SSB) phone and CW (Morse Code) on the 80m, 40m, 20m, 15m, and 10m bands. There were no WARC bands in 1972!

So, when my license arrived, I was ready to operate some phone. To be honest, I don't really remember my first phone contact, but after a few SSB contacts (or QSOs), I recall thinking, "Is that all there is?" The reality of finally being able to talk on the radio certainly did not live up to my expectations. I unplugged the microphone, connected up the key, and got back on Morse Code.

Why is it called CW? Quite often you'll hear the term CW when referring to Morse Code operation. CW is the abbreviation for "continuous wave," and refers to the mode of transmission, rather than the type of modulation (Morse Code) used to transmit information. The Random House Dictionary defines the term continuous wave as "an electromagnetic wave of constant amplitude and frequency." In a way, calling radiotelegraphy CW is a misnomer because the continuous wave is continually interrupted to form the dits and dahs of Morse Code, but hey, they didn't ask me when they started calling it CW.

I operated only sporadically through the rest of my high school years and my college years, and after college, I was off the air for many years. About twelve years ago, though, the ham radio bug hit me again.

On Field Day 2001, I wandered over to the local club's operation early Sunday Morning. There were only two guys there, one operating CW on 40m and the other guy eating breakfast. I knew the second guy from the bike club, so we got to talking.

He had been trying to make some contacts on 20m phone, but apparently the band conditions were pretty rough that morning, and he wasn't having much luck. Despite not being on the air for a long time, I suggested we try CW. He said that thought he'd seen a straight key around somewhere, and after digging through a couple of boxes, he actually did find one.

It didn't, of course, take long to find someone calling "CQ FD" short for "CQ Field Day." Both of us were so rusty, though, that it took a couple of tries at copying his call before we both agreed on his call sign. At that point, I said, "Go ahead. You can make the first contact." At that, he demurred, and said, "No. You go first."

With some hesitation, I tapped out our club call sign and waited. What do you know? He actually came back to us! What a thrill! It wasn't as thrilling as my first Novice contact, but it was pretty close.

We went on to make about a dozen CW contacts that morning, each one as painful, but as thrilling, as the first. Often, we had to listen multiple times to a station to get either the callsign or to be sure of the exchange information. Even so, it was great to be back on the air.

If you've read my blog (www.kb6nu.com) or my study guides

(www.kb6nu.com/tech-manual), you know how badly I've been hooked once again on amateur radio. Since that time, I've made more than 20,000 contacts—including contest contacts—and I'd guess that 95% or more of them have been Morse Code contacts.

Even my e-mail address reflects how I feel about Morse Code. I'm cwgeek@kb6nu.com.

I know that Morse Code isn't for everyone. Unlike a lot of old timers (OTs, sometimes called OFs, or *old farts*), I didn't curse the FCC for eliminating the Morse Code test. I know a lot of good hams that either just aren't interested in Morse Code, or for whatever reason, can't seem to learn it. I think it was a real shame that they were kept out of our hobby.

Having said that, I also think that it's a shame that a lot of hams don't try to learn and use Morse Code because they think that it's an outdated mode or that it will be too hard to learn. It's certainly not outdated, as you can hear any evening on the lower portions of the HF ham bands, and while it can be a challenge to learn, it's not rocket science, either.

If you're thinking about learning Morse Code, I hope this book will give you that final push to do so. If you already know Morse Code, I hope that this book will get you more active on CW. In either case, I hope that this book will help you have more fun with amateur radio.

73,

Dan KB6NU

DAN ROMANCHIK, KB6NU

LEARNING THE CODE

Of course, before you can make a CW contact, you are going to have to learn to send and receive Morse Code. This may seem like a daunting task, but remember, thousands of amateur radio operators and millions of commercial operators have learned to send and receive Morse Code. There's no reason you can't learn it, too.

The first step in learning Morse Code is simply to get started. You'll never learn it if you don't start.

If you have a PC, I suggest starting by downloading the G4FON CW Trainer from http://www.g4fon.net. There is no charge for the program. The program teaches Morse Code a character at a time, using what is called the Koch Method. The Koch Method teaches you to recognize characters by sound and not by counting the dots and dashes.

If you don't have a PC, but do have an Apple computer, or a CD player capable of playing MP3 files, you can obtain the K7QO Code Course from FISTS, the International Morse Preservation Society. This course is also free of charge. To get a copy, send $3.50 to FISTS, 478 CR 53 Rogersville, AL 35652-3503.

Is everbody 'appy?

These days, everyone seems to be using a cellphone or a tablet. You could play the .mp3 files from the K7QO Code Course on your phone or tablet, or you could download an app. I've never used any of the apps that are available today, so I asked my followers on Twitter what they would recommend. Here's what they had to say:

- Matthew Williams @W2MDW
  @kb6nu Ham Morse is the most flexible, feature packed.
  Morse Coach is simple & clean.
- David Pechey @KD2BMU
  @kb6nu I've been using Koch Trainer - $0.99 and Morse
  Code Trainer - Free. @W2MDW told me about Learn CW
  Online, which I really like.
- M0TEF - Alistair@M0TEF
  @kb6nu I really liked using dah dit on the iPhone for
  drilling through the alphabet as well as training modes. It
  has worked for me.
- Chris Kelling @n1wko
  @kb6nu I'm using "dah dit"
- g4tny @classicfibre
  @kb6nu I've used 'morse trainer' on iphone with some
  success curing my rustiness ;))
- Richard Daily@rdaily
  @kb6nu Morse-it, Ham Morse, CWSpeed and Codeman
  are good. Codeman is free.
- D. Robinson KK4PWE @DRobinson6268
  @kb6nu I second the Dah-Dit app on iPhone. That's all I've
  ever used and it's fun.

If you have a favorite, let me know, and I'll include it in a future
edition of this book

How do you get to Carnegie Hall?

There's an old joke that goes something like this: Two tourists are
walking around New York City, when they spot someone carrying
a music case. Thinking that the musician might know, they ask,
"How do you get to Carnegie Hall?" He replies, "Practice, practice,
practice."

It's the same way with Morse Code. Once you get one of these
apps or courses, you'll have to practice. Many hams advise
newcomers to practice daily, but not to overdo it. Too much
practice and you'll burn out. A good suggestion is to limit yourself
to two, fifteen-minute sessions per day.

What some people do is to use idle moments to go over the
sounds of Morse Code in their heads. Say that you're in your car
and you notice that the car ahead of you has the license plate
number ABC 123. Sound that out in your head (di-dah dah-di-di-

dit dah-di-dah-dit di-dah-dah-dah-dah di-di-dah-dah-dah di-di-di-dah-dah). You can use this technique with traffic signs, signs, and billboards.

Another suggestion is to use the "buddy system." Get a friend or spouse to learn the code with you. If you're a member of an amateur radio club, ask around and see if there are any other guys who'd like to learn with you. If you can, find an "Elmer" who is an experienced CW operator.

If no one in your club currently operates CW, consider joining the SolidCopyCW mailing list (http://groups.yahoo.com/group/SolidCpyCW/). On this list, you'll find many CW operators, including yours truly, who are willing and able to help out in any way they can. You could also join FISTS (http://www.fistsna.org) or one of the other CW clubs. These clubs have "code buddy" programs that pair you up with an experienced operator or someone like yourself who is just learning the code.

Another time-tested way to learn and get better at Morse Code is to listen to the W1AW code practice transmissions. W1AW sends excerpts from QST articles every weekday at 9 am, 4 pm, 7 pm, and 10 pm Eastern time. Code speeds vary from as low as 5 wpm up to 40 wpm.

The operating schedule for W1AW can be found at http://www.arrl.org/w1aw-operating-schedule. If you are unable to listen to W1AW code practice at the scheduled times, you can download .mp3 files by going to http://www.arrl.org/code-practice-files.

W1AW also sends out various bulletins, such as the DX bulletin in Morse Code that you can use for practice. Refer to the operating schedule for information as to when these bulletins are transmitted.

Don't do it

There are several code courses out there that purport to teach you the code by using various catchphrases that sound like the character. For example, one of the courses, uses the catchphrase "dog did it" for the letter D. That sounds very much like dah-di-dit, which is the sound for the letter D.

In general, most Morse Code teachers do not recommend learning the code this way. The reason for this is that while they are effective in learning the sounds of the letters and numbers, they are a hindrance when it comes to improving your code speed. The theory is that translating back and forth from the mnemonic to the actual character slows you down. You want to be able to recognize a character by its sound alone, not some crazy image that gets conjured up in your mind.

Stick with it

Don't be discouraged if it takes you a while to master the code. Learning the code has a steep learning curve, but if you stick with it, you'll master it. Also don't get discouraged if you don't copy 100%. Just as you don't need to hear every word when conversing with someone, you don't need to copy every single character to take part in a QSO.

If you do miss a character, just ignore it and listen for the next one. Don't let missing a character bog you down. If you like, you can write an underscore or just leave a space to denote a character that you missed, but even that's not really necessary. When you look over what you've written down, you'll be able to get the gist of what was sent by characters that you did copy correctly.

Learning to send

Learning to receive is by far the hardest part about learning Morse Code. When you start out, you'll be able to send much faster than you can receive, so learning to send is not a big deal when you're just starting out.

Even so, I think it's helpful to practice sending as you're learning to receive. I'm not a cognitive scientist, but I think that there's something about thinking about what sounds to make and then using a key to make those sounds that helps solidify that sound in the mind.

To practice your sending, you'll need a key and some kind of code practice oscillator to produce the sound. As for the key, I usually suggest that even newcomers use a paddle instead of a straight key. I'll write more about this in the chapter, "Choosing a Key," but the two biggest reasons that I suggest using a paddle is that you'll send better code right away with a paddle and using a paddle is easier on the arm and wrist.

To use a paddle, you'll need to have some kind of keyer. Most modern HF transceivers have built-in keyers and a way to disable the transmitter so you can use that rig as an expensive code practice oscillator. On Icom radios, for example, you set the break-in function to no break-in.

You can also use an external keyer for this. To use the keyer as a code practice oscillator, simply set it so that the internal speaker is enabled and the keying output is disabled, so that you don't key your transmitter while practicing. You can, of course, also unplug the cable connecting the keyer to your radio.

To get some feedback on how well you're sending, you could pipe the audio into a program like fldigi (http://www.w1hkj.com /Fldigi.html) or CW Skimmer (http://www.dxatlas.com/ cwskimmer/). These programs do a decent job of decoding CW, especially with a solid signal, and you can compare what the program receives with what you sent. Another way is to send to your "code buddy." If he or she can copy what you're sending, then you know you're doing a good job.

Ditch the pencil and paper to get faster

When operating Morse Code, there's always a debate over how fast one should go. Many hams are happy to plod along at 16 – 20 wpm, or sometimes even slower. The FISTS CW Club even uses the slogan, "Accuracy transcends speed." While there is some truth to this, There's no reason that you can't have both accuracy and speed, and I would encourage you to work at operating as fast as you can.

One reason for this is that at somewhere around 25 wpm, operating Morse Code becomes nearly as conversational as phone. And the more conversational a QSO is, the more fun it is for me. I can get beyond signal reports and equipment descriptions and actually learn something about the other operator or the town he or she lives in.

When I got back on the air, I was one of those operators who was stuck somewhere 15 – 16 wpm, but I wasn't happy about it. Then, I read an article in QST or CQ that the biggest obstacle to getting faster is copying on paper. According to the article, you can't really write any faster than 20 wpm, and most people can't even write that fast. So, if you insist on copying down each individual character, then the fastest that you'll be able to copy is 20 wpm.

That made sense to me, but also I think it's also a multitasking issue. I don't know about you, but I have a limited amount of brain power. If I have to use a portion of that brain power to write letters on paper, then it's brain power that I can't devote to decoding Morse Code.

After reading that article, I decided that I needed to learn to just copy in my head. I went cold turkey. I put away the pen and the paper, and aside from giving demos to visitors to my shack, I only copy code in my head.

I immediately started getting faster and can now copy at 35+ wpm. I don't know if the cold turkey method will work for you, but undoubtedly, the key to getting faster is to ditch the pencil and paper and start copying in your head. And remember: practice, practice, practice.

Stretch

Another way to get faster is to stretch. I don't mean getting up out of your chair and stretching your arms above your head (although it's not a bad idea to do that once in a while while operating Morse Code), but rather your code-copying muscles.

You do this by having a contact with someone who's sending just a little faster than what you're comfortable copying. With a little bit of concentration, you should be able to copy that operator and next time it will be a little easier.

Another tip for getting faster is to participate in contests. You don't have to get serious about winning just to participate. Often, I'll work a contest for a couple of hours just for the fun of it.

How does contesting help you get faster? The key is that in a contest what is sent is very well,-defined. For most contests, only call signs, a signal report (almost always "599"), and a state, ARRL section, or zone number. Because this information is so well-defined, it's easier to anticipate what is being sent, and you'll therefore be able to copy it more easily, even if it's being sent at a speed higher than what you can normally copy. It's just another way of stretching.

GETTING ON THE AIR

Another way to practice is to listen to Morse Code contacts on the ham bands. That's actually how I taught myself Morse Code. Not only will this improve your receiving ability, it will teach you how to conduct a Morse Code contact.

Tuning in

To make contacts, the first thing that you need to do is to learn how to tune your radio properly. When you tune in a signal, you want to make sure that you have tuned your radio to a frequency that is as close as possible to frequency on which the station you are receiving is transmitting. This process is often called "spotting" or "zero beating." To do this, you tune your radio so that the tone you hear is as close to the frequency of the sidetone–the tone that your transceiver generates when you transmit–as possible.

While most experienced operators simply do this by ear, some radios have a SPOT function that you can use for this purpose. What this function does is to inject a continuous tone into the audio output of the radio while you are tuning it. When you're right on frequency, the signal you're trying to tune in, will nearly disappear. If you're off frequency, you'll continue to hear the characters being sent.

Some radios, such as the Elecraft K3, have an automatic zero beat feature. To use this feature on a K3, you simply press the CWT button, and the right half of the S Meter display becomes a zero-beat tuning meter. As you tune in the signal, you can see if you are getting closer to the signal's frequency or farther away from it. Once you are close, you can tap the SPOT button, and the K3 will automatically zero beat the signal. How cool is that?

The method that I most often use with my older transceiver is to feed the audio of my radio into the laptop computer in my shack and view the signal with a program that produces an audio spectrograph, such as a PSK-31 program. I then simply tune the radio so that the audio frequency of the received signal matches the sidetone frequency. (I usually set my sidetone frequency to 600 Hz.) This is a very quick and easy way to zero beat a signal.

Sidetone frequency

The sidetone frequency is the frequency of the audio tone that your radio generates when you are sending. This will also be the frequency of the audio tone that you hear when your receive frequency matches the frequency of the station you are receiving. Personally, I prefer a 600 Hz tone, and usually set my sidetone frequency to this value. Some operators, though, prefer sidetones as low as 400 Hz and as high as 750 Hz. My suggestion is to play around with this setting and find the frequency at which a CW signal is most understandable.

Another thing to consider is that you may wish to vary the tone of the received signal during a QSO. This helps prevent fatigue, and can actually enhance the readability of the signal. Rather than actually change the sidetone setting, I do this by enabling Receiver Incremental Tuning (RIT) and then varying the RIT frequency to change the audio frequency of the received signal.

Filters

Another consideration when tuning in a CW signal is the choice of filters. In the past, radios would be equipped with fixed bandwidth filters, and for CW operation, 500 Hz was a popular choice. Some radios could be equipped with filters as narrow as 250 Hz.

Most modern radios use digital signal processing filters, and these filters can be set to almost any bandwidth. The reason for this is that the bandwidth is set in software and is not a hardware function at all. With this capability, you can set the bandwidth to exactly the value that you need for a particular contact. With the front panel controls, it is easy to set the bandwidth that you need.

I find that for most operating conditions, 700 Hz is the best setting for the receive bandwidth. Anything more and you hear stations 1 kHz above or below you. Anything less and the sound of the signal isn't as pleasing to my ear.

That doesn't mean that I never set the bandwith narrower than

700 Hz. When there is a nearby signal interfering with my contact, I'll narrow the bandwidth as much as I need to to eliminate the interference. A similar situation occurs when operating a contest. In a contest, there are many stations operating in close proximity, and in that situation, using a narrow bandwidth filter may be necessary.

Some radios have a feature called passband tuning. With this feature, you can not only set the width of of the filter, but also the center frequency of the filter. This is a very useful feature in the event that there is an interfering signal just above or just below your frequency. For example, if the interfering signal is just about your frequency, then you can adjust the upper cutoff frequency so that it is below the interfering signal and filter it out.

Making contact

Now that you know how to tune your radio properly, you are ready to make contact. You can do this in several ways: listen for someone calling CQ, call CQ yourself, wait until an ongoing contact ends and then call one of stations, or break into an ongoing contact.

When an operator wants to make a contact, he or she will "call CQ." What this means is that he or she is looking for a contact. A station calling CQ will send something like the following:

CQ CQ CQ CQ DE KB6NU KB6NU KB6NU K

CQ means exactly what it sounds like, that is "seek you." The K sent at the end of the call is a prosign that signifies that you've finished your transmission and that you're inviting other stations to call you now.

To reply to a station calling CQ, you send the other station's callsign, followed by "de," then your callsign. For example, if I heard W1ABC calling CQ, and I wanted to contact him, I would send:

W1ABC W1ABC W1ABC DE KB6NU KB6NU KB6NU K

If signals were very strong, I might send W1ABC's call only once or twice, and my call maybe only twice, but when replying to a CQ, you should always send both the calling station's callsign and your callsign. This procedure eliminates any confusion. The calling station knows that you are calling him, and repeating your callsign at least once helps the other stationbe sure that he received your callsign correctly.

BAD PRACTICE

It is becoming more common for stations responding to a CQ to send only their call, and to only send it a single time. While this is the norm when operating contests, or a DX pileup, it **IS NOT**, and **SHOULD NOT** be the norm, for most contacts. Unless you send the calling station's callsign, he cannot be sure that you are calling him, and if you don't repeat your callsign at least once, there's a good chance that he will not copy your call correctly. There's no reason to do this for most contacts, so just don't do it.

Different types of CQs

Sometimes you'll hear a station using what's called a "directional CQ." In addition to "CQ," the station will include a location or maybe a group name. For example, if W1ABC wanted to contact someone in New York, he might send:

CQ NY CQ NY CQ NY DE W1ABC W1ABC W1ABC K

This tells me that W1ABC wants to find someone in New York. Since I live in Michigan, I should not reply to that call

Sometimes, stations want to work only "DX" stations. DX stations are stations that are not in the same country, or in nearby countries, as the calling station. So, for example, if I hear:

CQ DX CQ DX CQ DX DE W1ABC W1ABC W1ABC K

I know not to reply to that CQ because W1ABC is looking to contact stations outside of the U.S. Chances are he doesn't want to talk to anyone in Canada either, as Canada borders the U.S. and is not really considered DX.

On the other hand, if you hear F1XYZ call CQ DX, feel free to answer. The prefix F1 signifies that the station is in France, and he would probably be glad to hear you call him.

Calling CQ

If you don't hear anyone calling CQ and you still want to make a contact, you can call CQ yourself. The first thing that you need to do is to find a clear frequency. Tune around and find a spot at least 1 kHz away from the nearest station. Then, before you actually send CQ, send the Q-signal **QRL?** QRL is a Q-signal that means, "Is this frequency in use?" (For a list of other Q-signals commonly used in amateur radio contacts, refer to the Reference and Resources section.)

If no one responds to your **QRL?** with **QRL** or **C** or **YES** or a callsign, you can then start calling CQ. Some books advise using the "3 x 3 x3" method. This consists of sending CQ three times, followed by sending your call three times, and repeating this entire process three times:

CQ CQ CQ DE KB6NU KB6NU KB6NU
CQ CQ CQ DE KB6NU KB6NU KB6NU
CQ CQ CQ DE KB6NU KB6NU KB6NU K

I find this to be entirely too long. Instead, I send CQ four times, followed by my callsign three times:

CQ CQ CQ CQ DE KB6NU KB6NU KB6NU K

After sending this, pause to listen for replies. Keep in mind that a station replying to your CQ may not be exactly on the same frequency that you transmitted on. Tune up and down a few kHz, listening for replies to your CQ.

Tail-ending

Another way to make contact with another station is "tail-ending." To tail-end a contact, simply wait until two stations that are conversing have finished their contact, then call one or the other. If the station you called can hear you, and is available for another contact, chances are he will reply to your call.

You should not call a station, if he or she ended his final transmission with the prosign **CL**. This prosign means that the station is closing down, or in other words going off the air.

Breaking in

Another way to make contact is to break into an existing contact. You would do this in much the same way that you break into a voice contact. That is, you would send the prosign BK or your callsign when one of the stations has turned it over to the other station. Then, wait for one of the stations already engaged in the contact to acknowledge you.

While this is a perfectly appropriate thing to do, it rarely happens in CW operation, certainly much less often than say 2m FM operation. I'm not exactly sure why this is the case. Perhaps one reason for this is that when most hams had separate transmitters and receivers, break-in operation was not a common practice, and this practice simply continued on to this day.

Whatever the reason, don't be surprised if you're not acknowledged and asked to join the contact should you try to break in. It's not that the two operators are shunning you. Since this is not a common practice, they may honestly not know how to respond.

What to send during a contact

Now that you've made contact, you need to know how and what to send to the other station during a contact. Most CW contacts follow a fairly simple protocol, at least on the first transmission. On the first transmission, most stations will send a signal report, the operator's name, and the station's location.

So, let's say that I've just called CQ, and W1ABC has just returned that call. Here is an example of what I might send on that first transmission:

W1ABC DE KB6NU TNX FER CALL—UR RST RST 599 599—NAME DAN DAN—QTH ANN ARBOR MI ? ANN ARBOR MI—HW CPI? W1ABC DE KB6NU K

Let's dissect this transmission. First, I send his call followed by my callsign. This confirms that I've received his callsign properly. If I've made a mistake, W1ABC can correct me when it's his turn to transmit.

Then, I thank the other operator for his call. We try to be as polite as we can on CW. This helps us avoid some of the childishness that sometimes goes on on the phone portions of the ham bands.

Next, I send a signal report using the RST format. See the References and Resources section for the meaning of each of the digits in an RST signal report. The signal report is followed by name and my location.

Note that I repeat the signal report, my name, and my location. It's always a good practice to repeat these. Even if signals are very strong at the beginning of a contact, signals might fade, or the other operator may be distracted and not copy my information correctly. Repeating them helps ensure that he gets the information.

When I sent my location, notice that I use the Q-signal QTH. Q-signals are three letter combinations that CW operators use in place of common phrases. In this case, QTH stands for "My location is _____." So, when I send **QTH ANN ARBOR MI** I am telling the other operator that I am located in Ann Arbor, MI.

Also note that I did not send the comma before MI. You will sometimes hear operators send the comma, but it's really not necessary. The transmission is perfectly understandable without it.

Between each sentence, I send the **BT** prosign instead of a period. The **BT** prosign is a long dash. There's no real reason for this, except that this is how CW operators do things. Sometimes, you will hear operators send the period, but in general, the use of **BT** is preferred.

Other prosigns that I used in this transmission include the **?** and **K**. The question mark means that I am going to report some important information. I could have also used it when I repeated the signal report and my name, but because those strings are so short, you rarely hear a **?** sent for those repeats. The **K** at the end of the transmission signifies that I am ending my transmission and inviting the other operator to begin his. A list of all the prosigns commonly used in CW contacts is found in the References and Resources section.

I also make use of a lot of CW abbreviations:

- DE—FROM
- TNX—THANKS
- FER—FOR
- UR—YOUR. Could also mean "you are," but not in ths case.
- HW—HOW
- CPI—COPY

The use of abbreviations is encouraged when operating CW. This reduces the number of characters that you must send, meaning that you can send more information in a shorter time. This enables a CW contact to proceed more like a normal conversation. You'll find a list of common CW abbreviations used on the ham bands in the References and Resources section.

W1ABC's turn

When W1ABC hears the **K** he knows that it's his turn to transmit. He might send the following:

**KB6NU DE W1ABC R TNX FER RPRT—UR RST RST
599 599—NAME JOE JOE—QTH BOSTON
MA ? BOSTON MA—BTU KB6NU DE W1ABC K**

This is very similar to my first transmission, but there are several differences. The **R** sent after his callsign denotes that he copied every thing I sent. Then, he's thanking me for the signal report, not my call. Finally, instead of **HW CPI?**, he sends **BTU**, which means "back to you."

In the second transmission, the stations often exchange descriptions of equipment and often a weather report. Being a technical hobby, I often enjoy hearing about what radios or what antennas the other station is using, but I find weather reports to be kind of boring, unless it's some unusual weather event. I usually skip the weather report and try to move the discussion in another direction. This is especially true in the winter, when I've contacted someone in Florida or Texas, where the weather is much warmer than it is here.

Really making contact

Now that the formalities are out of the way, you can *really* make contact. By that I mean really have a conversation with the other operator. There are some interesting people out there, but you'll never know that, if you don't make an effort to really talk to them.

One trick that I use to get the ball rolling sometimes is to look them up on QRZ.Com and read their profile. Often, operators will describe their activities and interests, not only in amateur radio, but outside of the hobby as well. For example, one time I looked up a guy, saw that he'd posted several photos of him on a bicycle, and since I've done some biking myself, we had a nice chat about bicycling.

Others will post pictures of their children and grandchildren, or even their pets. Asking about them is a great conversation starter, and you often hear great stories about them.

Even if they don't have any personal information on QRZ.Com, you can still have a great conversation. For example, if they live in a town with an interesting name, I'll ask them if they know the history of that name. One time, I worked a guy in Toad Suck, AR, and asked him about the origin of that name. That was a fun conversation.

There are many other ways to start a conversation, and I'd encourage you to do so. Just think about how you would start up a conversation with a stranger at a football game or a party or some other gathering. Ask them about what they do for a living or what got them interested in amateur radio or what projects that they're working on currently. If you do this, you'll be rewarded many times over.

Special events, contests, and working DX

There are occasions when really short contacts are more appropriate than long contacts. This is when making contact with special event stations and DX stations, or when participating in a contest. Short contacts are more appropriate in these cases, and there's even a different protocol for making contact.

To make contact with a DX station or a special event station that has many stations trying to contact him, you would normally only send your callsign a single time and then listen to see if the DX station heard you or not. If they did hear you, they will send your callsign, followed by a signal report (usually 599, whether you are actually 599 or not). In reply, you send **TU 599**. That ends the contact. If the DX station did not copy your callsign correctly you might also send your callsign again once or twice. This ensures that they've gotten it correctly.

In a contest, you generally have to also exchange some other kind of information. For example, in some contests, you have to also exchange your ARRL section. In others, you might have to exchange a serial number. Check the rules for each contest to make sure that you exchange the correct information.

Operating "split"

Sometimes you'll hear a DX or special even station append the word **UP** to their CQs. This means that they are "operating split" and want you to call them on a frequency higher than the frequency they are calling on. What this does is allows other stations to hear F1XYZ clearly, while at the same time allows F1XYZ to pick and choose the stations he wants to reply to. When there are many stations calling a single stations (this is often referred to as a "pileup"), it helps the DX station or special event station make as many contacts as possible.

For example, F1XYZ might send:

CQ DX CQ DX F1XYZ F1XYZ UP

That means he will be listening for calls at least 1 kHz above the frequency that he is on.

To contact this station, you will have to set up your radio to operate in split-frequency mode. That is, you'll have to set it up so that it transmits a kHz or two above the frequency on the main dial, which is the frequency of the DX station. You can use the SPLIT function or the XIT function of your radio to do this.

This chapter just scratches the surface of CW operating, but it should be enough to get you started. Getting on the air and practicing is the most important thing now. As you make contacts, you'll gain experience and confidence and become a better operator.

CHOOSING A KEY

When a new ham decides to learn Morse Code and start operating CW, one of the first things he or she must do is choose a key. There are many different types of keys available, and choosing one can be kind of confusing. With that in mind, let's look at the different keys that are available and discuss the pros and cons of each.

Straight key

The straight key is the most basic type of key. It has a single set of contacts, and the operator makes dits and dahs by holding down the key for different lengths of time. Because the design is so simple, this is usually the least expensive type of key you can purchase.

While many hams enjoy using straight keys, I'm not a big fan of them myself. It takes practice to make dits and dahs that are the same length over and over, and I guess that I just don't have the concentration necessary to do that.

Also, my arm tires very easily when using a straight key. I can't send very long before I begin to feel it in my wrist and forearm. Hams experienced with straight keys tell me that this is because I don't have the key adjusted properly or that I'm not holding my arm correctly. Whatever the reason, I can't really operate for more than 30 – 45 minutes with a straight key.

Many hams got their start in CW using a J-38 key. This key was made by several manufacturers during World War II, and there are many still in existence. This particular key was made by the Lionel Corporation, the same company that made Lionel trains.

Paddles

Paddles are keys that you use with an electronic keyer. They have two sets of contacts, one for the dits and one for the dahs. It doesn't matter how long that you hold down the key. Once a set of contacts is closed, the electronic keyer will make the dit or the dah. The nice thing about this arrangement is that the electronic keyer makes each dit and each dah the same length every time.

Another thing I like about the paddle is that it's very easy on the wrist and arm. Unlike the straight key, which you pump up and down, to operate a paddle, you rest your arm on the desk or table and simply actuate the paddle by moving your fingers or rotating your wrist. This is a lot less stressful, and I find that I can operate for hours using a paddle.

There are two main varieties of paddle: dual-lever and single-lever. The dual-lever paddle is sometimes called an iambic paddle. Both the single-lever and the dual-lever paddles have two sets of contacts, but in a single-lever paddle, the lever is common to both and only one set of contacts can be closed at a time. The dual-lever paddle has two completely-independent sets of contacts, and both can be closed simultaneously.

When both are closed, most electronic keyers will send alternating dits and dahs. This is called the iambic mode. More about how this works in the next chapter.

*There are two types of paddles: dual-lever (left)
and single-lever (right).*

Single-lever or dual lever paddle?

Chuck Adams, K7QO, has calculated that using a dual-lever paddle with an iambic keyer requires 11% fewer strokes than a single-lever paddle to send a message in Morse Code. Efficiency isn't the whole story, though. For one thing, it's easier to make mistakes with a dual-lever paddle. The reason for this is that the timing of the contact closures is critical when using a dual-lever paddle. If you make a contact too early or too late, or hold down a contact for too long, the code that the keyer will generate will be wrong. For example, instead of sending a C (dah-di-dah-dit), you end up sending dah-di-dah-di-dah.

This is one reason that the winners of high-speed CW contests tend to use single-lever paddles and not dual-lever paddles. They get penalized for making mistakes, and it's more difficult to make them with single-lever paddles.

You might also want to choose a single-lever paddle if you are used to using a semi-automatic key, or "bug." Using a single-lever paddle more closely resembles using a bug than does using a dual-lever paddle.

My recommendation is to try both and see which one you like best. Some operators will prefer the single-lever paddle for its simplicity, while others will prefer the dual-lever paddle.

Touch paddles

Several companies make "touch paddles." These paddles don't have levers, per se, but rather metal pads that one touches to close a contact. Instead of mechanically closing a contact, touch paddles have an integrated circuit them that senses the change in capacitance when you touch one of the pad then electronically closes a contact.

Many operators really like using touch paddles. Because there are no moving parts, there are no mechanical adjustments to make and no loud clicking sounds.

Semi-automatic keys, or "bugs"

Like the straight key, semi-automatic keys, or "bugs," are purely mechanical. The difference between a straight key and a bug, though, is that the bug has a mechanism that makes dits automatically. Dahs are still made manually, though.

Using one of these keys properly takes a lot of practice, and is generally not a good choice for a beginner. I have one myself, and although I only use it occasionally, I still haven't gotten the hang of using it after several years.

Computers and Keyboards

Since everything seems to be computerized, these days, why not sending CW? Some keyers have keyboard inputs, and many logging programs, such as N1MM and N3FJP, allow you to send CW via a computer comm port or USB interface. One advantage of using a keyboard is that, as long as you don't make any typos, you'll send perfect code. Sending via keyboard is also a boon to hams who may have arthritis and find operating a key or paddle by hand to be too painful.

Personally, I enjoy working a key by hand. There's something about doing this manually that appeals to me. I certainly don't have a problem with operators that use a keyboard, though. My feeling is that if using a keyboard gets them on the air, then more power to them. We need as many CW operators on the air as we can muster.

Which key is right for you?

If you do become a CW enthusiast, you'll find that you tend to collect keys and use them all from time to time. In my collection, I have:

- three straight keys, including the key I used as a Novice and a World War II-vintage J37 key with a leg clamp;

- four paddles, including three dual-lever and two single-lever paddles; and
- one semi-automatic key, also called a "bug."

My advice is to try them all and see which you like the best.

Adjusting and using your key

If my arguments here have failed to deter you from using a straight key, take some time to learn how to use one correctly. There's an excellent video (https://www.youtube.com/watch?v=DQj74Y2H8xQ), produced in 1966 by the Army's Office of the Chief Signal Officer, that shows you how to adjust and use a J-38 straight key. Most of this information is also applicable to other straight keys. For example, the video advises setting the gap between the two contacts to a distance equal to the thickness of four sheets of paper.

The video also includes advice on how to send using a straight key. It shows to position your arm, hand, and wrist to reduce strain and develop a good rhythm. It's all great advice, but personally, I'm sticking with a paddle and keyer.

Setting up a paddle is somewhat similar to a straight key, even though the lever moves from side to side instead of up and down. Whether you have a single-lever paddle or a dual-lever paddle, you'll have at least four adjustments to make: two contact adjustments and two spring tension adjustments.

If you're just starting out, you might want to adjust the contact spacing to be relatively wide, say the width of a dime. As you gain proficiency, you can always make this gap narrower. As far as spring tension goes, you need to make it stiff enough so that you don't make inadvertent errors. You don't, however, want to make this so stiff that moving the levers requires a lot of work.

Setting up and using a Vibroplex semi-automatic key, or "bug," can also be a challenge. Fortunately, there's an excellent video by Jim Wades, WB8SIW, that gives step-by-step instructions on how to set up a bug. He also gives some good advise on how to use a bug to send good code. You'll find this video at https://www.youtube.com/watch?v=qekmyx31Uxw.

Whichever key you use, feel free to experiment with the adjustments. I do this all the time. Sometimes, I prefer a narrow gap and a stiff spring tension. At other times, a loose feel is more to my liking. By experimenting, you can find the settings that best fit your sending style.

Key manufacturers/sellers

American Morse Equipment. www.americanmorse.com. American Morse sells both mini-paddles and regular-size paddles. For their Porta-Paddle, they even sell a leg mount, so you can operate portable or mobile.

Begali. www.i2rtf.com. When I used to bicycle, I only rode Italian bicycles. They not only were great bicycles, but they were stylish as well. That's how I feel about Begali keys. My favorite key is my Simplex, Begali's least expensive paddle, but I wish I could afford one of his fancier models.

Bencher. www.bencher.com. Bencher makes both inexpensive and more deluxe keys and paddles. I often recommend the Bencher BY-1 as a "starter" paddle. You can often find them online or at hamfests for $60-70.

Bulldog Keys. www.amateurradioproducts.com. Bulldog makes a link of small keys for QRP and portable operation enthusiasts.

CW Touch Paddles. www.cwtouchkeyer.com. This company only makes touch paddles.

Kent Engineers. www.kent-engineers.com. This company, located in England, has a long history in the Morse key business. They make straight keys, single-lever paddles, and dual-lever paddles (they call them single-paddle keys and twin-paddle keys). They are very nicely engineered and a good bargain.

Morse Express. www.morsex.com. Morse Express sells keys from Ameco (USA), Bencher (USA), BHC Bird Key (China), GHD (Japan), Hi-Mound (Japan), Palm Radio (Germany), Scheunemann Morsetasten (Germany), uniHAM (China), Vibroplex (USA), Nye Viking (USA), and MFJ (USA).

N3ZN Keys. www.n3znkeys.com. N3ZN makes some very nice hand-made keys.

Vibroplex. www.vibroplex.com. Vibroplex is the grand-daddy of key manufacturers in the U.S. They have been in business for more than 100 years. Their line of products includes straight keys, paddles, and semi-automatic keys, or "bugs."

There are more manufacturers and sellers out there, but I'll leave at that. If you have a favorite manufacturer that I haven't included here, please let me know, so that I can include them in a future edition.

KEYERS

In the previous chapter, I advise that even newcomers should learn to send with a paddle. If you send with a paddle, you will, of course, need an electronic keyer. An electronic keyer is a device that automatically forms the dits and the dahs when the operator closes one of the paddle contacts.

Most modern transceivers come with a built-in keyer, but many, if not most, CW operators prefer to use an outboard keyer. In general, outboard keyers are more flexible than built-in keyers. For example, some keyers have a USB interface that can be used to control them from logging and contesting programs. Only the latest transceivers have that capability.

Types of keying

Iambic A,B modes

As noted in the previous chapter, there are two kinds of paddles: single-lever and dual-lever. Dual-lever paddles allow you to send using one of the iambic modes. This arguably makes your sending more efficient. You cannot close both sets of contacts simultaneously using a single-lever paddle, so you can never operate in iambic mode with a single-lever paddle.

There are two iambic modes: mode A and mode B. Keyers using either of these modes will send alternating dits and dahs when both paddle contacts are closed simultaneously.

The difference between the two modes is what happens when both paddles are released. A keyer operating in mode A completes the element being sent when the operator releases the paddles. A keyer operating in mode B sends an additional elements opposite to the one being sent when the paddles are released.

Ultimatic mode

Ultimatic mode also uses dual-lever paddles, but operates differently than iambic mode. Instead of sending alternate dits and dahs, a keyer using the Ultimatic mode will continue to send the element of the contact last closed.

If you first closed, the dah contact and then the dit contact, an iambic key would send "dah di dah di dah..." An Ultimatic keyer, on the other hand, would send, "dah di di di di..."

The Ultimatic mode allows you to use both paddles for more characters than iambic mode. For example, to send "X," you:

- press the dah lever,
- press the dit lever, while holding down the dah lever for two dits, then
- release the dit lever for a single dah.

You can't do this operating in iambic mode.

I keep threatening to try the Ultimatic mode, but I never seem to get around to it. While it might offer some advantage over iambic mode, I'm not sure those advantages are big enough to go to the trouble of learning it.

Semi-automatic mode

Some keyers have a semi-automatic, or "bug" mode. In this mode, the keyer tries to simulate a bug. When you close the dit contact, it will send a series of dits automatically, but when you close the dah contact, the keyer will key for as long as you keep that contact closed.

Connecting the key to the keyer

Traditionally, we connect the contact operated by the thumb to the dit input of the keyer and the contact operated by the index finger to the dah input of the keyer. The reason for this is that's the way bugs work. You use the index finger to make dahs and the thumb to make dits.

This is just a convention, though, and there's no law that says you have to wire your keys this way. In fact, most keyers have the ability to use either connection for dits or dahs. I find that sticking to this convention to be convenient, though. When operating an event where there are multiple operators, such as Field Day or a special event station, you don't have to futz around with cables if everyone sticks to the convention.

Once you have that figured out, you have to decide what kind of connector to use. In the past, keys have connected to radios using ¼-in. phone plugs. The reason for this is that in the era of tube radios, there was a substantial amount of current flowing through that plug, and those phone plugs are designed to carry an amp or more.

In modern radios, however, keys connect to digital inputs and only a modest amount of current flows through the key and plug. That's one reason that most keyers nowadays use 1/8-in. (3.5 mm) phone jacks for key input connectors.

Connecting the keyer to the radio

If you're using an external keyer, you'll need a cable to connect the keyer to the straight key input of your radio. Many modern HF transceivers have two key jacks, one on the front panel and one on the rear panel. The jack on the front panel can usually be programmed to accept either a three-conductor, ¼-in. phone plug with the three connections from a paddle or a two-conductor, ¼-in. phone plug with the two connections from a straight key. The key jack on the rear panel is usually designed for a straight key and can normally accept only a two-conductor, ¼-in. phone plug. In my station, I find it most convenient to connect the keyer output to the ¼-in. phone jack on the rear panel of the transceiver.

Keyer manufacturers

Begali

Begali's keyer, called the CW Machine is more than just a keyer. It combines the functions of an iambic memory keyer and a keyboard keyer with the functionality of a logging program. Website: www.i2rtf.com/html/cw_machine.html

HamGadgets

HamGadgets sells a couple of inexpensive keyer kits: the PicoKeyer Plus ($19) and the Ultra PicoKeyer ($29). They are easy to build and work very well. One interesting feature of the HamGadgets keyers is that they have a modulated CW (MCW) mode. In this mode, the keying output can be connected to a transceiver's push-to-talk switch, while the audio is connected to the transceiver's microphone input. I have a PicoKeyer connected in this way to the 2m/440m rig in my shack, and every once in a while, I give folks on the repeater a shot of CW. Website: www.hamgadgets.com.

Hamcrafters

Hamcrafters sells a couple of different keyer kits, based on their own chip, the K12 keyer IC. I really like the WKUSB keyers and have built two of them. The kits are quite easy to build, and programs such as the N1MM contest logger and the N3FJP logging program easily interface with the keyers via the USB port. Website: www.hamcrafters.com.

MFJ

MFJ sells many different models of keyers, including some that have built-in paddles and some that are designed to be used with keyboards. Website: http://www.mfjenterprises.com/.

Idiom Press

Idiom Press offers two keyers: the LogiKey K-5 and the Super CMOS-4. They are both full-featured keyers and are available in kit form. Unfortunately, they lack a USB port, so they cannot be used with logging programs. Website: http://idiompress.com/keyers.php.

Unified Microsystems

The XT-4 from Unified Microsystems is a battery-powered CW memory keyer. The XT-4 is fully iambic with self-completing characters. The CW speed range is 8-45 WPM. Website: www.unifiedmicro.com/keyer.html

JUST HAVE FUN!

The most important thing about operating Morse Code is to just have fun. Now that knowing Morse Code is no longer a requirement, you can just relax and learn it and use it at you own pace. Find a key that you enjoy using and pound away at it. Make some contacts and get to know a whole new group of operators.

If you have any comments or questions, compliments or complaints, I want to hear from you. Please feel free to e-mail me at cwgeek@kb6nu.com. Of course, you can also find me down at the bottom of the bands. Look for me there and give me a call.

73,

Dan KB6NU

REFERENCES AND RESOURCES

RST Signal Reporting

A signal report is part of nearly every CW contact, and is normally a three-digit number, with the first digit denoting the readability (R), the second digit signal strength (S), and the third digit signal tone (T). A typical report that might be sent is "599" meaning the signal is perfectly readable, extremely strong, with a perfect tone. Below, find the meaning for each of the reports component parts.

Note that when giving a station a signal report for a phone transmission, the T is not included. T is only used for CW transmission signal reports.

Readability (R)
1—Unreadable
2—Barely readable, occasional words distinguishable
3—Readable with considerable difficulty
4—Readable with practically no difficulty
5—Perfectly readable

Signal Strength (S)
1—Faint signals, barely perceptible
2—Very weak signals
3—Weak signals
4—Fair signals
5—Fairly good signals
6—Good signals

7—Moderately strong signals
8—Strong signals
9—Extremely strong signals

Tone (T)

1—Sixty cycle a.c. or less, very rough and broad
2—Very rough a.c., very harsh and broad
3—Rough a.c. tone, rectified but not filtered
4—Rough note, some trace of filtering
5—Filtered rectified a.c. but strongly ripple-modulated
6—Filtered tone, definite trace of ripple modulation
7—Near pure tone, trace of ripple modulation
8—Near perfect tone, slight trace of modulation
9—Perfect tone, no trace of ripple or modulation of any kind

Additional characters

Other characters are sometimes added after the RST to denote some other characteristic of the transmission:

C—Chirp. Added to denote if the signal is chirping, that is rapidly changing frequency.

K—Clicks. This character is added to denote that the other stations signals has "key clicks." These are spurious emissions that occur when a CW signal's risetime is too fast. These clicks can interfere with nearby QSOs.

Q-Signals

QRG _____ / **QRG?** - Your frequency is _____. / What's my exact frequency?

QRL / QRL? - I am busy. / Are you busy?
The most common usage for this Q-signal is to ask if a frequency is in use before beginning to call CQ or another station. The station wishing to use the frequency would send "QRL?" If the frequency is in use, the station using the frequency, should send "QRL" or simply "C."

QRM / QRM? - You are being interfered with. / Is my transmission being interfered with?
QRM is often used (incorrectly) as a noun. For example, someone might send, "LOTS OF QRM TONITE."

QRN / QRN? - You are being troubled by static or atmospheric noise. / Are you troubled by static or atmospheric noise?
Like QRM, QRN is often misused as a noun. For example, someone might send, "LOTS OF QRN ON THE BAND TONITE."

QRO / QRO? - I will increase power. / Shall I increase transmitter power?
Often, QRO is used as an adjective. For example, someone might send, "I AM QRO NW," meaning that his transmitter is running a significant amount of power, usually more than the 100 W, typical of today's transceivers.

QRP / QRP? - I will decrease power. / Shall I decrease transmitter power?
QRP is most often used as an adjective. If someone says, "RIG HR IS QRP," what they mean is that the power output of the transmitter is 5 W or less. To qualify for the QRP category of most contests your transmitter output power must be 5 W or less.

QRQ / QRQ? - Send faster. / Shall I send faster?

QRS / QRS? - Send slower. / Shall I send slower?
Feel free to use this Q-signal should you get into a contact with someone who is sending faster than you can receive. The courteous thing to do is for the faster operator to slow down when requested to do so.

QRT / QRT? - Stop sending. / Shall I stop sending?

Common amateur usage is a little different than the original meaning. When someone sends "MUST QRT" or "WILL QRT ON NXT XMSN" they mean that they are going to go off the air.

QRU / QRU? - I have nothing for you. / Do you have anything for me?
Sometimes you'll hear stations say, "I AM QRU." What this means is that unless you have something further, they wish to end the contact. They probably also mean this if they send "QRU?"

QRV / QRV? - I am ready. / Are you ready?
Originally, this meant that a station was ready to copy a message. Nowadays, it means that they are ready to get on the air.

QRZ _____ / QRZ? - _____ is calling you. / Who is calling me?
In contests or DX operation, a station will often send "QRZ?" to denote that he's finished with one contact and will begin listening for other stations. This is a little different than the original meaning.

QSB / QSB? - Your signals are fading. / Are my signals fading?
Like QRM and QRN, QSB is often used as a noun (instead of the word "fading") even sometimes as an adjective. For example, someone might say, "THE BAND IS VY QSB TONITE."

QSK / QSK? - I can work break-in. / Can you work break-in?
Stations that have break-in capability switch rapidly from transmit to receive even between individual dits and dahs. That allows the receiving station to "break in" in the middle of a transmission. When this happens, the transmitting station should stop sending to allow the receiving station to make a comment. In practice, few operators actually do this, though.

QSL / QSL? - I acknowledge receipt. / Can you acknowledge receipt?
QSL was originally meant to be used to acknowledge receipt of a formal message. Now, it is often used to denote that a transmission was received, whether or not it contained a formal message. We also use it as an adjective, as in "QSL card." A QSL card acknowledges that we had a contact with the station to whom we sent the card.

QSO____ / QSO____? - I can communicate with ____ directly. / Can you communicate with ____ directly?
QSO was often used in the early days of amateur radio when the range of a station was limited and stations relayed messages from

one to another. Nowadays, we mostly use QSO as a noun, meaning a contact with another station.

QSY / QSY? - I will change frequency. / Shall I change frequency?

QTH / QTH? - My location is. / What is your location?
You often hear QTH used as a noun. People often send "QTH IS XXXX." While not an egregious use of this Q-signal, it is incorrect.

CW Abbreviations

ABT: About
ADDR: Address
AGN: Again
AM: Amplitude Modulation, Morning
ANT: Antenna
BCI: Broadcast Interference
BCNU: Be seeing you
BK: Back, Break, Break in
BN: Been
BTR: Better
BTWN: Between
BUG: Semi-Automatic key
BURO: Bureau (QSL bureau)
BUX: Bucks, as in dollars
B4: Before
C: Yes, Correct
CFM: Confirm; I confirm
CK: Check
CKT: Circuit
CL: Call
CLDY: Cloudy
CLG: Calling
CLR: Clear
CONDX: Conditions
CPI: Copy
CPY: Copy
CQ: Calling any station
CU: See YouCUD: Could
CUL: See You later
CUM: Come
CUZ: Because
CW: Continuous wave
DE: From, This Is
DIFF: Difference
DN: Down
DR: Dear
DX: Distance
EL: Element
ES: And
EVE: evening

FB: Fine Business, excellent
FER: For
FM: Frequency Modulation: From
FREQ: Frequency
GA: Go ahead, Good Afternoon
GB: Good bye, God Bless
GE: Good Evening
GESS: Guess
GG: Going
GM: Good morning
GN: Good night
GND: Ground
GUD: Good
HI: The telegraph laugh
HPE: Hope
HQ: Headquarters
HR: Here; Hear
HRD: Heard
HV: Have
HW: How, How Copy?
LID: A poor operator
LTR: Later; letter
LV: Leave
LVG: Leaving
LW: Long wire
MA: Milliamperes
MNI: Many
MORN: Morning
MSG: Message
NR: Number
NW: Now
OB: Old boy
OC: Old chap
OM: Old man
OP: Operator
OPR: Operator
OT: Old timer; Old top
OVR: Over
PKG: Package
PM: Afternoon, evening
PROB: Problem
PSE: Please
PT: Point

PWR: Power
R: Received as transmitted; Are; Decimal Point
RCV: Receive
RCVR: Receiver
REF: Refer to; Referring to; Reference
RFI: Radio frequency interference
RIG: Station equipment
RPT: Repeat, Report
RPRT: Report
RTTY: Radio teletype
SED: Said
SEZ: Says
SIG: Signature; Signal
SKED: Schedule
SN: Soon
SRI: Sorry
SSB: Single Side Band
STN: Station
SUM: Some
SW: Switch, Shortwave
SWL: Short wave listener
TEST: Contest
THRU: Through
TIL: Until
TMW: Tomorrow
TMRW: Tomorrow
TKS: Thanks
TNX: Thanks
TR: Transmit
TRBL: Trouble
T/R: Transmit/Receive
TT: That
TU: Thank you
TVI: Television interference
TX: Transmitter; Transmit
TXT: Text
U: You
UR: Your; You're
URS: Yours
VFO: Variable Frequency Oscillator
VY: Very
WKD: Worked
WKG: Working

WL: Well; Will
WUD: Would
WX- Weather
XCVR: Transceiver
XMIT: Transmit
XMSN: Transmission
XMTR: Transmitter
XTAL: Crystal
XYL: Wife
YL: Young lady
YR: Year
72: Low power version of 73
73: Best Regards
88: Love and kisses

Procedural signs, or prosigns

Prosigns are combinations of characters (although a couple are just a single character) that signify that a particular point in a contact has been reached or that call for an action by the receiving operator. Although the prosigns are written below as two separate characters, you should sending them in a single sequence. For example, the prosign **AR** should be sent as di-dah-di-dah-dit.

AR
End of message. You would send AR when you're done sending a message and are ready to turn it over to the other operator. AR is sent before you send callsigns. For example, at the end of your first transmission, you might send:

....HW? AR W8ABC DE KB6NU K

In practice, however, this is rarely done, but you will hear it from time to time.

AS
Stand by. You would send AS if you want the other station to stand by for a short time while you tweak your antenna tuner or perhaps make a note in your log.

BK
Break. Send BK when you want the other station to start transmitting without going through the station identification process. Feel free to make liberal use of this prosign. Remember that you really only have to identify your station once every ten minutes. You also use BK when beginning a transmission that's responding to an invitation to break in.

CL
Closing. This signifies that you are going to be going off the air. You send CL at the end of the station ID on your very last transmission of a QSO. For example:

...73 ES GN SK W8ABC DE KB6NU CL

K, KN
Go ahead. You send either K or KN after identifying your station to

invite the other station to start transmitting. If you are open to allowing other stations to join your conversation, send K. If you want only the station you are currently in contact with to go ahead, send KN.

R

All received OK. Use this prosign near the beginning of your transmission to signify that you received what was sent. I usually take this to mean that the other station copied my transmission solidly, or in other words, copied every single character.

SK

End of message. SK is similar to AR, but means that this will be the last transmission of a contact.

VE

Understood. This prosign is not used often, but you do hear it occasionally.

CW Clubs

FISTS. The FISTS Club, (International Morse Preservation Society) was founded in 1987 by the late George "Geo" Longden G3ZQS of Darwen, Lancashire England, after recognizing a need for a club in which veteran operators would help newcomers and less-experienced operators learn and improve CW proficiency. During the first year, membership reached 300, most of whom were in Great Britain and Europe. Nancy Cott, WZ8C, formed the North American Chapter in 1990, and was the president of the group until sadly she became a Silent Key in 2014. There are also FISTS chapters in Australia and Asia.

FISTS sponsors a variety of activities, including the CW Buddy program, operating activities, including a quarterly sprint contest, and operating awards. Dues in the U.S. are $15/year. For more information, visit www.fists.co.uk, www.fistsna.org, www.fistsdownunder.org, and www.feacw.net.

North American QRP CW Club (NAQCC). The NAQCC is similar to FISTS in that they exist to promote the use of CW, and sponsor a number of operating activities and awards, but differ in that they also promote the use of low-power (QRP) operation and simple wire antennas. There are no dues to join NAQCC. For more informations, visit www.naqcc.info.

Straight Key Century Club (SKCC). Formed in 2006, the SKCC's focus is on the use of mechanical keys, including both straight keys and semi-automatic keys or "bugs." Like FISTS and NAQCC, they offer a variety of operating activities and awards, and like NAQCC, membership is free. For more information, visit www.skccgroup.com.

CW Operator's Club (CWOps). CWops was launched in January 2010, and has since grown from a handful of members to nearly 1,000 members today in more than 70 countries. They sponsor several contests and the CW Academy that teaches newcomers and veteran hams the art of CW using a novel "virtual" training environment that allows an instructor to work with hams located thousands of miles apart. Dues are $12/yr, and you must be sponsored by current CWOps members. For more information, go to www.cwops.org.

First-Class Operator's Club (FOC). Founded in 1938 in the UK, the First Class CW Operators' Club (FOC) promotes good CW (Morse code) operating, activity, friendship and socializing. Its membership is currently about 500. To join the FOC, you must be able to operate at 25 wpm or more and be nominated by a current member, and then be sponsored by four current members. Their website notes that you, "may feel free to ask members for information about the club, actually asking to be nominated is frowned upon." For more information, go to www.g4foc.org.

Books, Websites & Mailing Lists

Books

The Art and Skill of Radio-Telegraphy (http://morse-rss-news.sourceforge.net/artskill.pdf. http://www.zerobeat.net/tasrt/index.html, http://www.zerobeat.net/tasrt/contents.htm)

Zen and the Art of Radio-Telegraphy

Morse Code: Breaking the Barrier

Morse Code: The Essential Language

Websites

http://www.radiotelegraphy.net/

http://www.skccgroup.com/member_services/link_library/

Mailing Lists

Solid Copy CW (http://groups.yahoo.com/group/SolidCpyCW/) (Yahoo Group)

Vibroplex Collector's Assn. (http://groups.yahoo.com/group/VibroplexCollectorsAssn/) (Yahoo Group)

ABOUT THE AUTHOR

I have been a ham radio operator since 1971 and a radio enthusiast as long as I can remember. In addition to being an active CW operator on the HF bands:

- I blog about amateur radio at KB6NU.Com (http://kb6nu.com), one of the leading amateur radio blogs on the Internet.
- I have written a series of study guides for the Technician Class, General Class, and Extra Class exams. You can find the *No-Nonsense Technician Class License Study Guide, No-Nonsense General Class License Study Guide,* and the *No-Nonsense Extra Class License Study Guide* in PDF, Nook (ePub) and Kindle (Mobipocket) formats on my website at http://www.kb6nu.com/study-guides/. The Kindle versions are available on Amazon, while the Nook versions are available on Barnes&Noble.
- I am the author of *21 Things to Do With your Amateur Radio License*, an e-book for those who have been recently licensed or just getting back into the hobby. This book is also available on my website (http://www.kb6nu.com/shop/) and Amazon and Barnes&Noble.
- I send out a monthly column to approximately 350 amateur radio clubs throughout North America for publication in their newsletters.
- I am the station manager for WA2HOM (http://www.wa2hom.org), the amateur radio station at the Ann Arbor Hands-On Museum (http://www.aahom.org).
- I teach amateur radio classes around the state of Michigan.

- I serve as the ARRL Michigan Section Training Manager and conduct amateur radio leadership workshops for amateur radio club leaders in Michigan.

You can contact me by sending e-mail to cwgeek@kb6nu.com. If you have comments or question about any of the stuff in this book, I hope you will do so.

29633854R00030

Made in the USA
Columbia, SC
23 October 2018